Student Name:

Pre-Test Score:

Date Started:

Date Completed:

Final Test Score:

DIVISION

4th Grade

SAREEDO

Pre-University MATH

WORKBOOKS

Are You Ready 1?

1)
$$\begin{array}{r} 12 \\ +56 \\ \hline \end{array}$$
$$\begin{array}{r} 27 \\ -25 \\ \hline \end{array}$$
$$\begin{array}{r} 17 \\ +68 \\ \hline \end{array}$$
$$\begin{array}{r} 49 \\ +79 \\ \hline \end{array}$$

2)
$$\begin{array}{r} 26 \\ +84 \\ \hline \end{array}$$
$$\begin{array}{r} 30 \\ -21 \\ \hline \end{array}$$
$$\begin{array}{r} 91 \\ -58 \\ \hline \end{array}$$
$$\begin{array}{r} 37 \\ +59 \\ \hline \end{array}$$

3)
$$\begin{array}{r} 45 \\ -36 \\ \hline \end{array}$$
$$\begin{array}{r} 47 \\ +59 \\ \hline \end{array}$$
$$\begin{array}{r} 74 \\ -38 \\ \hline \end{array}$$
$$\begin{array}{r} 39 \\ +59 \\ \hline \end{array}$$

4)
$$\begin{array}{r} 52 \\ +52 \\ \hline \end{array}$$
$$\begin{array}{r} 54 \\ -54 \\ \hline \end{array}$$
$$\begin{array}{r} 67 \\ +67 \\ \hline \end{array}$$
$$\begin{array}{r} 45 \\ -37 \\ \hline \end{array}$$

5)
$$\begin{array}{r} 129 \\ +346 \\ \hline \end{array}$$
$$\begin{array}{r} 194 \\ +271 \\ \hline \end{array}$$
$$\begin{array}{r} 786 \\ -588 \\ \hline \end{array}$$
$$\begin{array}{r} 616 \\ -497 \\ \hline \end{array}$$

Are You Ready 2?

1)
$$312 + 860$$
$$245 + 257$$
$$875 - 683$$

2)
$$152 + 967$$
$$524 - 258$$
$$917 - 686$$

3)
$$712 - 565$$
$$274 + 256$$
$$176 + 688$$

4)
$$600 - 506$$
$$494 + 546$$
$$774 - 675$$

5)
$$5712 + 5659$$
$$2014 - 1567$$

Multiplication Review 1

	a	b	c	d
1)	936 × 2	439 × 3	514 × 4	937 × 5
2)	967 × 6	468 × 7	535 × 8	915 × 9
3)	908 × 7	407 × 5	506 × 4	906 × 5
4)	904 × 3	406 × 4	508 × 6	
5)	736 × 32	745 × 58	756 × 42	

Multiplication Review 2

	a	b	c	d
1)	324 × 32	327 × 23	334 × 14	324 × 35
2)	404 × 46	454 × 57	427 × 48	405 × 59
3)	320 × 37	412 × 25	303 × 54	450 × 65
4)	308 × 436	270 × 617	406 × 528	

Multiplication Review: Extra Practice

$$\begin{array}{r} 324 \\ \times\,132 \\ \hline \end{array} \qquad \begin{array}{r} 327 \\ \times\,523 \\ \hline \end{array} \qquad \begin{array}{r} 334 \\ \times\,654 \\ \hline \end{array} \qquad \begin{array}{r} 425 \\ \times\,235 \\ \hline \end{array}$$

$$\begin{array}{r} 404 \\ \times\,456 \\ \hline \end{array} \qquad \begin{array}{r} 454 \\ \times\,650 \\ \hline \end{array} \qquad \begin{array}{r} 407 \\ \times\,408 \\ \hline \end{array} \qquad \begin{array}{r} 405 \\ \times\,507 \\ \hline \end{array}$$

$$\begin{array}{r} 320 \\ \times\,900 \\ \hline \end{array} \qquad \begin{array}{r} 602 \\ \times\,685 \\ \hline \end{array} \qquad \begin{array}{r} 703 \\ \times\,254 \\ \hline \end{array}$$

Divide

1) $2 \div 2 =$

2) $14 \div 7 =$

3) $24 \div 4 =$

4) $4 \div 2 =$

5) $24 \div 6 =$

6) $30 \div 6 =$

7) $8 \div 2 =$

8) $81 \div 9 =$

9) $81 \div 3 =$

10) $6 \div 2 =$

11) $54 \div 6 =$

12) $20 \div 5 =$

13) $10 \div 2 =$

14) $63 \div 9 =$

15) $10 \div 5 =$

16) $14 \div 2 =$

17) $40 \div 8 =$

18) $40 \div 8 =$

19) $16 \div 2 =$

20) $36 \div 9 =$

21) $36 \div 4 =$

22) $12 \div 2 =$

23) $64 \div 8 =$

24) $12 \div 4 =$

25) $18 \div 2 =$

26) $28 \div 7 =$

27) $18 \div 6 =$

28) $20 \div 2 =$

29) $42 \div 6 =$

30) $42 \div 7 =$

Divide

1) $22 \div 2 =$

2) $24 \div 8 =$

3) $26 \div 4 =$

4) $42 \div 2 =$

5) $25 \div 6 =$

6) $30 \div 5 =$

7) $84 \div 2 =$

8) $85 \div 9 =$

9) $51 \div 3 =$

10) $62 \div 2 =$

11) $55 \div 5 =$

12) $21 \div 7 =$

13) $19 \div 2 =$

14) $65 \div 7 =$

15) $19 \div 5 =$

16) $15 \div 2 =$

17) $45 \div 8 =$

18) $32 \div 8 =$

19) $17 \div 2 =$

20) $37 \div 9 =$

21) $27 \div 3 =$

22) $12 \div 5 =$

23) $65 \div 8 =$

24) $48 \div 6 =$

25) $28 \div 5 =$

26) $28 \div 6 =$

27) $38 \div 6 =$

28) $26 \div 5 =$

29) $47 \div 5 =$

30) $43 \div 7 =$

Divide

1) $2\overline{)2}$

8) $2\overline{)4}$

15) $2\overline{)6}$

22) $6\overline{)60}$

2) $2\overline{)8}$

9) $2\overline{)10}$

16) $2\overline{)12}$

23) $3\overline{)24}$

3) $2\overline{)14}$

10) $2\overline{)16}$

17) $2\overline{)18}$

24) $3\overline{)29}$

4) $2\overline{)20}$

11) $2\overline{)22}$

18) $2\overline{)24}$

25) $3\overline{)28}$

5) $3\overline{)3}$

12) $3\overline{)6}$

19) $3\overline{)19}$

6) $3\overline{)12}$

13) $3\overline{)15}$

20) $3\overline{)18}$

7) $3\overline{)24}$

14) $3\overline{)27}$

21) $3\overline{)30}$

Divide

1) $4\overline{)8}$

2) $4\overline{)20}$

3) $4\overline{)32}$

4) $4\overline{)44}$

5) $4\overline{)24}$

6) $3\overline{)33}$

7) $3\overline{)24}$

8) $4\overline{)12}$

9) $4\overline{)24}$

10) $4\overline{)36}$

11) $4\overline{)48}$

12) $3\overline{)27}$

13) $3\overline{)36}$

14) $3\overline{)27}$

15) $4\overline{)16}$

16) $4\overline{)30}$

17) $4\overline{)40}$

18) $3\overline{)21}$

19) $2\overline{)16}$

20) $4\overline{)48}$

21) $3\overline{)30}$

22) $4\overline{)50}$

23) $3\overline{)42}$

24) $3\overline{)65}$

25) $3\overline{)53}$

Divide:

1) $5\overline{)0}$

2) $5\overline{)15}$

3) $5\overline{)30}$

4) $5\overline{)45}$

5) $5\overline{)60}$

6) $6\overline{)18}$

7) $6\overline{)36}$

8) $5\overline{)5}$

9) $5\overline{)20}$

10) $5\overline{)35}$

11) $5\overline{)50}$

12) $6\overline{)6}$

13) $6\overline{)24}$

14) $6\overline{)42}$

15) $5\overline{)10}$

16) $5\overline{)25}$

17) $5\overline{)40}$

18) $5\overline{)55}$

19) $6\overline{)12}$

20) $6\overline{)30}$

21) $6\overline{)48}$

22) $5\overline{)43}$

23) $5\overline{)27}$

24) $6\overline{)38}$

25) $6\overline{)59}$

Divide

1) $6\overline{)54}$

2) $6\overline{)66}$

3) $7\overline{)21}$

4) $7\overline{)42}$

5) $7\overline{)63}$

6) $7\overline{)84}$

7) $5\overline{)45}$

8) $7\overline{)27}$

9) $7\overline{)57}$

10) $7\overline{)28}$

11) $7\overline{)49}$

12) $7\overline{)70}$

13) $6\overline{)54}$

14) $5\overline{)45}$

15) $6\overline{)63}$

16) $7\overline{)14}$

17) $7\overline{)35}$

18) $7\overline{)56}$

19) $7\overline{)77}$

20) $6\overline{)42}$

21) $7\overline{)28}$

22) $7\overline{)43}$

23) $6\overline{)57}$

24) $6\overline{)78}$

25) $5\overline{)59}$

Divide

1) $8\overline{)0}$

2) $8\overline{)24}$

3) $8\overline{)48}$

4) $8\overline{)72}$

5) $8\overline{)96}$

6) $9\overline{)18}$

7) $9\overline{)45}$

8) $8\overline{)8}$

9) $8\overline{)32}$

10) $8\overline{)56}$

11) $8\overline{)80}$

12) $9\overline{)0}$

13) $9\overline{)27}$

14) $9\overline{)54}$

15) $8\overline{)16}$

16) $8\overline{)40}$

17) $8\overline{)64}$

18) $8\overline{)88}$

19) $9\overline{)9}$

20) $9\overline{)36}$

21) $9\overline{)63}$

22) $4\overline{)43}$

23) $5\overline{)51}$

24) $6\overline{)68}$

25) $7\overline{)73}$

Divide

1) $9\overline{)72}$

2) $9\overline{)99}$

3) $10\overline{)10}$

4) $10\overline{)40}$

5) $10\overline{)70}$

6) $10\overline{)100}$

7) $9\overline{)72}$

8) $9\overline{)81}$

9) $9\overline{)98}$

10) $10\overline{)20}$

11) $10\overline{)50}$

12) $10\overline{)80}$

13) $10\overline{)107}$

14) $8\overline{)56}$

15) $9\overline{)90}$

16) $10\overline{)0}$

17) $10\overline{)30}$

18) $10\overline{)60}$

19) $10\overline{)90}$

20) $10\overline{)120}$

21) $8\overline{)72}$

22) $6\overline{)72}$

23) $8\overline{)71}$

24) $9\overline{)69}$

25) $7\overline{)84}$

Divide

1) $11\overline{)0}$ $11\overline{)11}$ $11\overline{)22}$

2) $11\overline{)33}$ $11\overline{)44}$ $11\overline{)55}$

3) $11\overline{)66}$ $11\overline{)77}$ $11\overline{)88}$

4) $11\overline{)99}$ $11\overline{)110}$ $11\overline{)120}$

5) $12\overline{)0}$ $12\overline{)12}$ $12\overline{)24}$

6) $12\overline{)36}$ $12\overline{)48}$ $12\overline{)60}$

7) $12\overline{)72}$ $12\overline{)84}$ $12\overline{)96}$

8) $12\overline{)108}$ $12\overline{)120}$ $9\overline{)90}$

9) $9\overline{)81}$ $4\overline{)24}$ $7\overline{)56}$

10) $8\overline{)32}$ $6\overline{)36}$ $5\overline{)30}$

11) $8\overline{)40}$ $7\overline{)63}$ $9\overline{)63}$

12) $6\overline{)48}$ $8\overline{)48}$ $9\overline{)72}$

13) $8\overline{)72}$ $8\overline{)40}$ $8\overline{)24}$

14) $6\overline{)24}$ $8\overline{)56}$ $7\overline{)56}$

Division Facts (Timed Review-1)

1) $2\overline{)2}$ $3\overline{)6}$ $4\overline{)12}$

2) $6\overline{)18}$ $3\overline{)18}$ $2\overline{)12}$

3) $4\overline{)24}$ $5\overline{)25}$ $2\overline{)18}$

4) $3\overline{)24}$ $6\overline{)24}$ $3\overline{)27}$

5) $3\overline{)21}$ $6\overline{)36}$ $5\overline{)45}$

6) $3\overline{)15}$ $4\overline{)36}$ $4\overline{)32}$

7) $3\overline{)30}$ $6\overline{)72}$ $5\overline{)35}$

8) $7\overline{)35}$ $8\overline{)32}$ $9\overline{)36}$

9) $7\overline{)49}$ $8\overline{)40}$ $9\overline{)72}$

10) $9\overline{)27}$ $8\overline{)24}$ $7\overline{)21}$

11) $7\overline{)42}$ $7\overline{)56}$ $8\overline{)56}$

12) $9\overline{)45}$ $8\overline{)48}$ $7\overline{)28}$

13) $8\overline{)32}$ $9\overline{)27}$ $5\overline{)30}$

14) $9\overline{)36}$ $8\overline{)40}$ $9\overline{)72}$

Division Facts (Timed Review-2)

1) $9 \overline{)9}$ $8 \overline{)16}$ $7 \overline{)14}$

8) $7 \overline{)35}$ $8 \overline{)32}$ $9 \overline{)36}$

2) $6 \overline{)18}$ $9 \overline{)18}$ $2 \overline{)18}$

9) $7 \overline{)56}$ $8 \overline{)40}$ $9 \overline{)72}$

3) $3 \overline{)24}$ $4 \overline{)24}$ $6 \overline{)24}$

10) $9 \overline{)81}$ $8 \overline{)48}$ $7 \overline{)21}$

4) $7 \overline{)42}$ $8 \overline{)24}$ $3 \overline{)27}$

11) $4 \overline{)32}$ $8 \overline{)56}$ $9 \overline{)18}$

5) $8 \overline{)48}$ $6 \overline{)66}$ $5 \overline{)40}$

12) $9 \overline{)45}$ $8 \overline{)48}$ $7 \overline{)28}$

6) $3 \overline{)15}$ $4 \overline{)36}$ $4 \overline{)32}$

13) $8 \overline{)32}$ $9 \overline{)27}$ $5 \overline{)30}$

7) $3 \overline{)21}$ $6 \overline{)72}$ $5 \overline{)35}$

14) $9 \overline{)36}$ $7 \overline{)63}$ $9 \overline{)72}$

Divide

1) $2\overline{)82}$

2) $6\overline{)72}$

3) $2\overline{)48}$

4) $6\overline{)84}$

5) $2\overline{)96}$

6) $5\overline{)80}$

7) $9\overline{)45}$

8) $3\overline{)48}$

9) $7\overline{)84}$

10) $3\overline{)54}$

11) $7\overline{)98}$

12) $3\overline{)87}$

13) $6\overline{)96}$

14) $9\overline{)54}$

15) $4\overline{)56}$

16) $8\overline{)96}$

17) $4\overline{)64}$

18) $8\overline{)88}$

19) $4\overline{)96}$

20) $2\overline{)96}$

21) $9\overline{)63}$

22) $5\overline{)65}$

23) $9\overline{)99}$

24) $5\overline{)75}$

25) $7\overline{)77}$

Divide

1) $2\overline{)54}$

2) $6\overline{)96}$

3) $2\overline{)68}$

4) $6\overline{)60}$

5) $2\overline{)80}$

6) $5\overline{)77}$

7) $7\overline{)49}$

8) $3\overline{)63}$

9) $2\overline{)84}$

10) $3\overline{)60}$

11) $7\overline{)70}$

12) $3\overline{)90}$

13) $6\overline{)68}$

14) $7\overline{)56}$

15) $4\overline{)60}$

16) $3\overline{)96}$

17) $4\overline{)80}$

18) $8\overline{)80}$

19) $4\overline{)49}$

20) $2\overline{)25}$

21) $7\overline{)63}$

22) $5\overline{)60}$

23) $4\overline{)98}$

24) $5\overline{)50}$

25) $9\overline{)90}$

Divide:

1) $2\overline{)52}$

2) $6\overline{)76}$

3) $2\overline{)49}$

4) $6\overline{)67}$

5) $2\overline{)87}$

6) $5\overline{)85}$

7) $8\overline{)48}$

8) $3\overline{)65}$

9) $2\overline{)85}$

10) $3\overline{)65}$

11) $7\overline{)79}$

12) $3\overline{)70}$

13) $6\overline{)66}$

14) $8\overline{)56}$

15) $4\overline{)61}$

16) $3\overline{)43}$

17) $4\overline{)86}$

18) $8\overline{)89}$

19) $4\overline{)40}$

20) $2\overline{)26}$

21) $8\overline{)64}$

22) $5\overline{)61}$

23) $4\overline{)65}$

24) $5\overline{)58}$

25) $8\overline{)90}$

Divide:

1) $\dfrac{11R2}{4\overline{)4\ 6}}$

2) $3\overline{)3\ 4}$

3) $3\overline{)3\ 8}$

4) $3\overline{)6\ 7}$

5) $3\overline{)6\ 8}$

6) $2\overline{)5\ 3}$

7) $2\overline{)6\ 7}$

8) $2\overline{)4\ 9}$

9) $4\overline{)4\ 7}$

10) $5\overline{)4\ 7}$

11) $6\overline{)4\ 7}$

12) $7\overline{)4\ 7}$

13) $8\overline{)4\ 7}$

14) $9\overline{)6\ 7}$

15) $2\overline{)5\ 3}$

16) $3\overline{)5\ 3}$

17) $2\overline{)6\ 9}$

18) $4\overline{)4\ 9}$

19) $4\overline{)8\ 5}$

20) $5\overline{)4\ 9}$

21) $3\overline{)6\ 7}$

22) $2\overline{)3\ 3}$

23) $2\overline{)4\ 3}$

24) $5\overline{)4\ 8}$

25) $3\overline{)4\ 7}$

Divide:

1)
$$2 \overline{)248}$$
$$124$$

6)
$$2 \overline{)648}$$

11)
$$2 \overline{)226}$$

16)
$$2 \overline{)824}$$

2)
$$3 \overline{)663}$$

7)
$$3 \overline{)960}$$

12)
$$2 \overline{)628}$$

17)
$$3 \overline{)393}$$

3)
$$2 \overline{)626}$$

8)
$$2 \overline{)224}$$

13)
$$3 \overline{)936}$$

18)
$$2 \overline{)120}$$

4)
$$4 \overline{)448}$$

9)
$$2 \overline{)844}$$

14)
$$4 \overline{)840}$$

19)
$$3 \overline{)651}$$

5)
$$2 \overline{)472}$$

10)
$$3 \overline{)657}$$

15)
$$2 \overline{)456}$$

Divide:

1) 3)651

6) 5)160

11) 3)126

16) 2)128

2) 2)120

7) 8)120

12) 3)129

17) 4)148

3) 3)120

8) 5)150

13) 3)141

18) 2)100

4) 4)120

9) 5)140

14) 3)162

5) 6)120

10) 5)130

15) 3)198

Divide:

1) 4)2 5 2

2) 4)2 2 0

3) 4)2 2 8

4) 4)2 3 2

5) 4)2 3 6

6) 2)1 6 0

7) 6)2 2 0

8) 6)2 6 4

9) 6)1 9 8

10) 6)1 4 4

11) 3)1 0 5

12) 3)2 0 7

13) 3)2 8 8

14) 3)3 0 6

15) 3)1 6 8

16) 2)1 3 8

17) 4)1 2 8

18) 2)1 5 0

Divide:

1) $3\overline{)252}$ 6) $7\overline{)168}$ 11) $4\overline{)104}$ 16) $2\overline{)614}$

2) $5\overline{)220}$ 7) $7\overline{)203}$ 12) $5\overline{)325}$ 17) $2\overline{)218}$

3) $3\overline{)228}$ 8) $7\overline{)630}$ 13) $5\overline{)345}$ 18) $2\overline{)416}$

4) $3\overline{)231}$ 9) $7\overline{)791}$ 14) $6\overline{)306}$

5) $3\overline{)231}$ 10) $6\overline{)462}$ 15) $6\overline{)156}$

Divide:

1) $3\overline{)3\ 2\ 1}$

2) $5\overline{)5\ 2\ 0}$

3) $3\overline{)6\ 2\ 4}$

4) $3\overline{)9\ 1\ 2}$

5) $3\overline{)2\ 3\ 1}$

6) $7\overline{)7\ 4\ 2}$

7) $7\overline{)7\ 0\ 7}$

8) $7\overline{)7\ 3\ 5}$

9) $4\overline{)4\ 1\ 6}$

10) $6\overline{)6\ 1\ 2}$

11) $4\overline{)4\ 0\ 8}$

12) $2\overline{)2\ 1\ 4}$

13) $5\overline{)3\ 5\ 5}$

14) $6\overline{)3\ 8\ 4}$

15) $6\overline{)6\ 5\ 4}$

16) $4\overline{)4\ 1\ 2}$

17) $5\overline{)5\ 1\ 5}$

18) $6\overline{)6\ 3\ 6}$

Divide:

1) $8 \overline{)256}$ 　 6) $9 \overline{)261}$ 　 11) $9 \overline{)936}$ 　 16) $7 \overline{)136}$

2) $8 \overline{)824}$ 　 7) $8 \overline{)464}$ 　 12) $3 \overline{)309}$ 　 17) $5 \overline{)515}$

3) $8 \overline{)608}$ 　 8) $9 \overline{)639}$ 　 13) $4 \overline{)288}$ 　 18) $2 \overline{)150}$

4) $9 \overline{)189}$ 　 9) $9 \overline{)630}$ 　 14) $9 \overline{)171}$

5) $6 \overline{)234}$ 　 10) $8 \overline{)144}$ 　 15) $8 \overline{)816}$

Divide:

1) 2)6 3 0   6) 7)6 3 0   11) 4)6 3 6   16) 4)5 7 6

2) 3)6 1 5   7) 8)8 4 0   12) 2)2 1 4   17) 5)4 5 0

3) 4)6 3 2   8) 9)6 3 0   13) 5)5 2 5   18) 6)3 6 0

4) 5)6 3 0   9) 9)6 5 7   14) 6)3 3 0

5) 6)6 3 0   10) 6)6 1 8   15) 6)6 5 4

Divide:

1) $3\overline{)861}$

2) $2\overline{)820}$

3) $3\overline{)873}$

4) $4\overline{)532}$

5) $6\overline{)330}$

6) $5\overline{)860}$

7) $9\overline{)126}$

8) $5\overline{)750}$

9) $5\overline{)840}$

10) $5\overline{)125}$

11) $3\overline{)124}$

12) $3\overline{)525}$

13) $8\overline{)400}$

14) $7\overline{)700}$

15) $3\overline{)600}$

16) $2\overline{)700}$

17) $4\overline{)412}$

18) $2\overline{)516}$

Divide:

1)
$$6\,2\text{R}2$$
$$4\,)\,2\ 4\ 6$$

6)
$$7\,)\,7\ 4\ 5$$

11)
$$4\,)\,4\ 0\ 9$$

16)
$$4\,)\,4\ 1\ 3$$

2)
$$2\,)\,5\ 3\ 1$$

7)
$$7\,)\,7\ 0\ 8$$

12)
$$2\,)\,8\ 1\ 8$$

17)
$$5\,)\,5\ 1\ 8$$

3)
$$2\,)\,3\ 4\ 5$$

8)
$$7\,)\,7\ 3\ 9$$

13)
$$5\,)\,3\ 5\ 7$$

18)
$$6\,)\,6\ 3\ 9$$

4)
$$3\,)\,9\ 1\ 3$$

9)
$$4\,)\,4\ 1\ 8$$

14)
$$6\,)\,3\ 8\ 1$$

5)
$$3\,)\,2\ 3\ 3$$

10)
$$6\,)\,6\ 1\ 5$$

15)
$$6\,)\,6\ 5\ 7$$

Divide:

1) 8)2 5 7

6) 9)2 6 0

11) 3)9 3 6

16) 6)1 3 5

2) 8)8 2 7

7) 8)4 6 6

12) 4)3 0 9

17) 5)5 1 8

3) 8)6 0 9

8) 9)9 3 9

13) 5)2 8 8

18) 3)1 5 7

4) 9)1 0 9

9) 2)6 3 1

14) 6)1 7 1

5) 6)2 3 5

10) 7)1 4 9

15) 4)7 1 5

Divide:

1) $2\overline{)631}$

2) $3\overline{)616}$

3) $4\overline{)633}$

4) $5\overline{)633}$

5) $6\overline{)635}$

6) $7\overline{)639}$

7) $8\overline{)842}$

8) $9\overline{)937}$

9) $6\overline{)657}$

10) $6\overline{)519}$

11) $4\overline{)437}$

12) $2\overline{)215}$

13) $5\overline{)629}$

14) $6\overline{)633}$

15) $7\overline{)654}$

16) $8\overline{)576}$

17) $5\overline{)256}$

18) $3\overline{)310}$

2 Minutes Practice: Complete the Tables By filling the blanks:

	Total	Each Group is:	Number of Groups
1)	40	8	5
2)	56		7
3)	90	2	
4)	45		15
5)	60	4	
6)	80		8
7)	120		6
8)	320	4	
9)	156		13
10)	165	15	
11)	210	15	
12)	240		17
13)	816		51
14)	756	63	
15)	891		11

2 Minutes Practice: Complete by choosing the missing number:

1)   **2, 6, 4, 10** $10 \div 2 = 5$

2)   **12, 6, 4, 10**   ____ $\div$ __4__ $= 3$

3)   **24 , 6, 48, 3**   ____ $\div$ ____ $= 8$

4)   **32, 16, 4, 8**   ____ $\div$ ____ $= 4$

5)   **52, 56, 4, 7**   ____ $\div$ ____ $= 8$

6)   **64, 16, 24, 48**   ____ $\div$ ____ $= 2$

7)   **2, 600, 5, 200**   ____ $\div$ ____ $= 120$

8)   **24, 8, 4, 72**   ____ $\div$ ____ $= 9$

9)   **8, 6, 40, 50**   ____ $\div$ ____ $= 5$

10)   **5, 7, 35, 70**   ____ $\div$ ____ $= 10$

11)   **11, 121, 4, 14**   ____ $\div$ ____ $= 11$

You worked hard. It is now time To

What is your dream?

_____
_____
_____
_____

what are the things that you are good at?

_____
_____
_____
_____

What is one thing in your life you like to improve?

_____
_____
_____
_____

Divide: Follow and understand these examples:

```
        5  3   R 1
   5 | 2  6  6
      -2  5  ↓
          1  6
         -1  5
```

```
        4  4   R 2
   6 | 2  6  6
      -2  4  ↓
          2  6
         -2  4
```

```
        3  8   R 0
   7 | 2  6  6
      -2  1  ↓
          5  6
         -5  6
```

```
        3  3   R 1
   8 | 2  6  6
      -2  4  ↓
          2  6
         -2  4
```

```
        2  9   R 5
   9 | 2  6  6
      -1  8  ↓
          8  6
         -8  1
```

```
        8  8   R 2
   3 | 2  6  6
      -2  4  ↓
          2  6
         -2  4
```

Divide:

	a	b	c	d

1) 5) 2 5 6   6) 2 5 7   7) 2 9 6   8) 2 5 8

2) 2) 3 1 6   3) 3 2 6   4) 4 0 6   5) 4 0 6

3) 6) 5 5 6   7) 6 5 6   8) 6 3 6   9) 6 2 6

4) 2) 7 1 6   3) 7 2 6   4) 8 3 6   5) 5 4 6

5) 5) 9 5 6   6) 9 6 6   7) 3 7 6   8) 4 8 6

Divide:

	a	b	c	d
1)	5) 2 4 6	6) 1 5 7	7) 2 3 6	8) 2 5 9
2)	2) 3 1 5	3) 3 2 5	4) 4 0 7	5) 4 0 8
3)	6) 3 4 6	7) 3 4 6	8) 3 3 6	9) 5 0 1
4)	2) 5 0 1	3) 5 0 1	4) 5 0 1	5) 5 0 1
5)	5) 4 1 7	6) 4 1 7	7) 4 1 7	8) 4 1 7

Divide:

	a	b	c	d

1) 5 ⟌ 5 1 9   6 ⟌ 5 1 9   7 ⟌ 5 1 9   8 ⟌ 5 5 6

2) 2 ⟌ 3 1 7   3 ⟌ 4 2 6   4 ⟌ 5 0 6   5 ⟌ 6 0 6

3) 6 ⟌ 7 5 6   7 ⟌ 2 5 6   8 ⟌ 3 3 6   9 ⟌ 4 2 6

4) 6 ⟌ 5 1 6   3 ⟌ 6 2 7   4 ⟌ 8 3 7   5 ⟌ 5 4 7

5) 9 ⟌ 9 5 8   6 ⟌ 9 6 7   7 ⟌ 3 7 9   8 ⟌ 4 8 5

Divide:

	a	b	c	d

1)   9) 9 0 7   8) 9 0 7   7) 9 0 7   6) 9 5 9

2)   5) 9 1 5   4) 9 0 7   3) 9 0 7   2) 9 0 8

3)   9) 8 4 6   8) 7 4 6   7) 6 3 6   6) 5 0 3

4)   5) 4 0 1   4) 3 0 1   3) 2 0 7   2) 1 0 5

5)   9) 2 1 7   8) 3 1 7   3) 3 1 7   2) 5 1 7

Divide:

	a	b	c	d

1) 3 | 2 0 6   4 | 1 5 7   5 | 2 3 6   6 | 2 5 9

2) 7 | 3 1 5   8 | 3 2 5   9 | 4 0 7   3 | 4 0 8

3) 4 | 3 4 6   5 | 3 4 6   6 | 3 3 6   2 | 5 6 1

4) 2 | 5 7 1   3 | 5 8 1   4 | 5 9 1   5 | 5 0 3

5) 5 | 4 1 7   6 | 4 1 7   7 | 4 1 7   8 | 4 1 7

Divide:

	a	b	c	d

1) 9 ⟌ 9 0 7   8 ⟌ 9 0 7   7 ⟌ 9 0 7   6 ⟌ 9 5 9

2) 5 ⟌ 9 1 5   4 ⟌ 9 0 7   3 ⟌ 9 0 7   2 ⟌ 9 0 8

3) 9 ⟌ 8 4 6   8 ⟌ 7 4 6   7 ⟌ 6 3 6   6 ⟌ 5 0 3

4) 5 ⟌ 4 0 1   4 ⟌ 3 0 1   3 ⟌ 2 0 7   2 ⟌ 1 0 5

5) 9 ⟌ 2 1 7   8 ⟌ 3 1 7   3 ⟌ 3 1 7   2 ⟌ 5 1 7

Divide:

	a	b	c	d
1)	9) 7 2 0	8) 6 4 0	7) 6 3 0	6) 3 6 0
2)	5) 4 5 0	4) 2 4 0	3) 1 2 0	3) 9 0 0
3)	9) 9 0 0	8) 8 0 0	7) 6 3 0	6) 4 2 0
4)	5) 4 0 0	4) 4 0 1	4) 8 0 1	2) 1 0 1
5)	7) 2 1 0	2) 2 0 0	3) 3 0 0	2) 5 0 0

Divide:

	a	b	c

1)

$2 \overline{)1\ 2\ 4\ 8}$ $\qquad$ $3 \overline{)9\ 0\ 7\ 5}$ $\qquad$ $4 \overline{)4\ 0\ 2\ 4}$

2)

$5 \overline{)1\ 9\ 0\ 5}$ $\qquad$ $6 \overline{)9\ 0\ 7\ 6}$ $\qquad$ $7 \overline{)9\ 0\ 7\ 4}$

3)

$8 \overline{)3\ 4\ 5\ 6}$ $\qquad$ $4 \overline{)1\ 4\ 8\ 4}$ $\qquad$ $5 \overline{)8\ 0\ 7\ 0}$

4)

$9 \overline{)9\ 0\ 1\ 8}$ $\qquad$ $8 \overline{)9\ 0\ 7\ 6}$ $\qquad$ $2 \overline{)5\ 0\ 6\ 8}$

Divide:

	a	b	c

1) $3 \overline{)5\ 6\ 0\ 1}$ $4 \overline{)6\ 0\ 4\ 2}$ $5 \overline{)4\ 6\ 0\ 3}$

2) $6 \overline{)5\ 3\ 0\ 4}$ $7 \overline{)8\ 0\ 7\ 5}$ $8 \overline{)9\ 1\ 0\ 6}$

3) $9 \overline{)3\ 0\ 5\ 7}$ $6 \overline{)1\ 8\ 0\ 8}$ $5 \overline{)8\ 0\ 7\ 9}$

4) $3 \overline{)1\ 0\ 9\ 3}$ $4 \overline{)9\ 0\ 7\ 4}$ $2 \overline{)6\ 0\ 6\ 7}$

Divide:

	a	b	c
1)	9 ⟌ 1 2 4 7	8 ⟌ 9 0 7 5	7 ⟌ 4 0 2 4
2)	6 ⟌ 1 9 0 8	5 ⟌ 9 0 7 6	4 ⟌ 9 0 7 4
3)	3 ⟌ 3 4 5 9	2 ⟌ 1 4 8 4	3 ⟌ 8 0 7 0
4)	4 ⟌ 9 0 1 6	5 ⟌ 9 0 7 6	6 ⟌ 5 0 6 8

Divide:

	a	b	c

1)   7 | 9 0 4 9   6 | 8 0 7 6   5 | 7 1 0 7

2)   4 | 6 9 0 8   3 | 5 0 7 7   2 | 4 0 7 5

3)   2 | 1 4 5 3   9 | 3 4 8 5   8 | 2 0 7 2

4)   6 | 9 0 0 7   4 | 9 0 7 7   5 | 5 0 0 9

You worked hard. Time to Relax and color. Yeah!

Use four five colors or more. Let's see what you get!

Divide: Follow and understand these examples:

```
      1  R1              2 R 0              3 R 6
15 | 2  6          13 | 2  6          15 | 5 1
   -1  5              -2  6                -4 5
   ――――              ――――               ――――
      1
```

```
   5 2 R1             3 1 R5             2 1 R3
15 | 7  8  0       25 | 7  8  0       28 | 5  9  1
   -7 5 ↓             -7 5 ↓             -5 6 ↓
   ――――              ――――               ――――
      3  0              3  0               3  1
      3  0              2  5               2  8
```

Divide:

	a	b	c

1)

$$12\overline{)49}$$ $$12\overline{)27}$$ $$12\overline{)37}$$

2)

$$12\overline{)51}$$ $$12\overline{)63}$$ $$12\overline{)57}$$

3)

$$12\overline{)84}$$ $$12\overline{)74}$$ $$12\overline{)36}$$

4)

$$12\overline{)40}$$ $$12\overline{)30}$$ $$12\overline{)20}$$

Divide:

	a	b	c

1) 13 ⟌ 3 9   13 ⟌ 5 7   13 ⟌ 3 7

2) 13 ⟌ 5 7   13 ⟌ 6 5   13 ⟌ 5 9

3) 13 ⟌ 8 0   13 ⟌ 8 4   13 ⟌ 3 7

4) 13 ⟌ 4 9   13 ⟌ 3 6   13 ⟌ 6 0

Divide:

	a	b	c
1)	15 ⟌ 5 5	15 ⟌ 3 7	15 ⟌ 5 7
2)	15 ⟌ 7 1	15 ⟌ 8 3	15 ⟌ 9 7
3)	15 ⟌ 2 4	15 ⟌ 3 4	15 ⟌ 4 6
4)	15 ⟌ 5 0	15 ⟌ 6 0	15 ⟌ 7 0

Divide:

	a	b	c
1)	17 ⟌ 3 7	17 ⟌ 4 7	17 ⟌ 3 9
2)	17 ⟌ 5 6	17 ⟌ 6 9	17 ⟌ 5 6
3)	17 ⟌ 7 4	17 ⟌ 8 4	17 ⟌ 3 5
4)	17 ⟌ 4 1	17 ⟌ 3 5	17 ⟌ 2 1

Divide:

	a	b	c
1)	19 ⟌ 3 8	19 ⟌ 3 8	19 ⟌ 5 7
2)	19 ⟌ 5 8	19 ⟌ 8 3	19 ⟌ 7 7
3)	19 ⟌ 8 5	19 ⟌ 7 6	19 ⟌ 5 6
4)	19 ⟌ 4 8	19 ⟌ 4 0	19 ⟌ 5 0

Divide:

	a	b	c

1)

$21\overline{)89}$ $21\overline{)77}$ $21\overline{)67}$

2)

$21\overline{)57}$ $21\overline{)73}$ $21\overline{)67}$

3)

$21\overline{)94}$ $21\overline{)84}$ $21\overline{)46}$

4)

$21\overline{)50}$ $21\overline{)60}$ $21\overline{)70}$

Divide:

	a	b	c

1)

2 3 | 7 9   2 3 | 6 7   2 3 | 5 7

2)

2 3 | 4 1   2 3 | 5 3   2 3 | 6 7

3)

2 3 | 7 4   2 3 | 8 4   2 3 | 4 6

4)

2 3 | 5 0   2 3 | 4 0   2 3 | 6 0

Divide:

	a	b	c

1)

$25\overline{)49}$ $25\overline{)27}$ $25\overline{)37}$

2)

$24\overline{)51}$ $24\overline{)63}$ $24\overline{)57}$

3)

$25\overline{)84}$ $25\overline{)74}$ $26\overline{)36}$

4)

$20\overline{)40}$ $25\overline{)30}$ $25\overline{)20}$

Divide:

	a	b	c

1)
$$22\overline{)502}\qquad 22\overline{)904}\qquad 22\overline{)807}$$

2)
$$22\overline{)815}\qquad 22\overline{)908}\qquad 22\overline{)907}$$

3)
$$21\overline{)846}\qquad 21\overline{)746}\qquad 21\overline{)636}$$

4)
$$21\overline{)401}\qquad 22\overline{)301}\qquad 22\overline{)242}$$

Divide:

	a	b	c
1)	12) 9 0 7	12) 6 0 6	12) 8 0 4
2)	12) 9 1 5	12) 9 0 7	12) 9 8 6
3)	12) 4 4 6	12) 5 4 6	12) 4 3 6
4)	12) 5 0 1	12) 6 0 1	12) 7 0 8

Divide:

	a	b	c

1)   22 | 8 0 7   27 | 9 2 7   27 | 6 0 7

2)   29 | 8 1 4   27 | 4 0 7   22 | 5 0 7

3)   21 | 8 4 6   21 | 7 4 2   21 | 6 3 3

4)   24 | 4 0 4   24 | 3 0 5   26 | 2 0 7

5)   26 | 4 0 1   26 | 3 0 1   26 | 2 8 7

Divide:

	a	b	c
1)	28) 5 8 8	28) 4 2 6	29) 6 0 0
2)	29) 9 1 4	28) 5 2 7	28) 6 3 5
3)	28) 8 5 6	28) 7 6 6	29) 6 6 8
4)	29) 4 3 8	29) 3 0 2	29) 8 0 3
5)	22) 4 4 1	22) 3 6 1	22) 2 8 7

Divide:

	a	b	c

1) $21\,\overline{)8\ 0\ 7}$ $22\,\overline{)9\ 2\ 7}$ $23\,\overline{)6\ 0\ 7}$

2) $24\,\overline{)9\ 1\ 5}$ $25\,\overline{)4\ 0\ 7}$ $26\,\overline{)5\ 0\ 7}$

3) $27\,\overline{)8\ 4\ 6}$ $28\,\overline{)7\ 4\ 6}$ $29\,\overline{)6\ 3\ 6}$

4) $11\,\overline{)4\ 0\ 1}$ $12\,\overline{)3\ 0\ 1}$ $13\,\overline{)2\ 0\ 4}$

5) $14\,\overline{)4\ 0\ 1}$ $15\,\overline{)3\ 0\ 1}$ $16\,\overline{)2\ 0\ 5}$

Divide:

	a	b	c

1)   30 ⟌ 6 0 5   31 ⟌ 9 2 7   32 ⟌ 4 0 8

2)   30 ⟌ 8 1 5   31 ⟌ 4 0 7   32 ⟌ 5 0 7

3)   30 ⟌ 7 4 6   31 ⟌ 8 4 6   32 ⟌ 8 3 6

4)   30 ⟌ 4 0 8   31 ⟌ 4 0 1   32 ⟌ 5 0 7

5)   30 ⟌ 4 0 3   31 ⟌ 3 0 2   32 ⟌ 2 0 4

Divide:

	a	b	c

1)   33 | 7 9 2   32 | 9 0 7   34 | 9 0 7

2)   33 | 9 1 5   35 | 9 0 7   35 | 9 0 7

3)   36 | 5 4 6   36 | 7 4 6   43 | 6 3 6

4)   41 | 4 0 1   42 | 3 0 1   42 | 2 0 7

5)   43 | 8 0 1   42 | 7 0 1   42 | 6 0 7

Divide:

	a	b	c
1)	37 $\overline{\smash{)}827}$	37 $\overline{\smash{)}927}$	39 $\overline{\smash{)}909}$
2)	39 $\overline{\smash{)}615}$	39 $\overline{\smash{)}707}$	39 $\overline{\smash{)}407}$
3)	37 $\overline{\smash{)}546}$	37 $\overline{\smash{)}546}$	33 $\overline{\smash{)}736}$
4)	38 $\overline{\smash{)}421}$	38 $\overline{\smash{)}331}$	38 $\overline{\smash{)}227}$
5)	33 $\overline{\smash{)}881}$	33 $\overline{\smash{)}771}$	33 $\overline{\smash{)}693}$

Divide:

	a	b	c
1)	41 ⟌ 9 0 7	42 ⟌ 4 4 4	42 ⟌ 5 5 5
2)	42 ⟌ 9 1 5	42 ⟌ 9 0 7	52 ⟌ 6 6 7
3)	41 ⟌ 8 4 6	41 ⟌ 9 0 6	41 ⟌ 6 3 6
4)	41 ⟌ 4 0 1	42 ⟌ 3 0 1	42 ⟌ 2 0 7
5)	51 ⟌ 5 5 1	52 ⟌ 7 6 5	52 ⟌ 8 3 2

Divide:

	a	b	c

1)   45 ⟌ 5 8 5   45 ⟌ 6 0 5   45 ⟌ 3 0 6

2)   37 ⟌ 9 1 5   37 ⟌ 9 0 3   37 ⟌ 5 0 7

3)   45 ⟌ 8 4 6   45 ⟌ 7 4 0   55 ⟌ 7 1 6

4)   45 ⟌ 8 5 0   45 ⟌ 7 0 5   45 ⟌ 3 0 7

5)   55 ⟌ 9 0 0   55 ⟌ 6 0 5   55 ⟌ 8 2 5

Divide:

	a	b	c

1)

51) 2 2 7   57) 3 3 7   62) 4 4 7

2)

62) 5 5 5   62) 6 5 7   62) 9 0 7

3)

61) 8 4 6   61) 7 4 6   63) 6 3 6

4)

61) 4 0 1   62) 3 0 1   62) 2 0 7

5)

63) 8 0 1   63) 7 0 1   63) 7 5 6

Divide:

	a	b	c

1)   65) 9 7 5   67) 9 8 7   67) 9 9 7

2)   66) 9 7 9   66) 9 8 7   65) 9 2 7

3)   66) 8 4 6   67) 7 4 6   67) 9 3 6

4)   65) 4 0 1   65) 3 0 1   65) 2 0 7

5)   64) 8 0 1   64) 7 0 1   64) 8 3 2

Divide:

	a	b	c

1)   68 | 9 9 4   68 | 8 8 7   69 | 9 7 7

2)   66 | 9 0 5   71 | 9 0 9   75 | 8 2 5

3)   72 | 7 9 6   73 | 8 4 6   60 | 8 8 6

4)   75 | 4 0 1   71 | 3 0 1   55 | 2 0 7

5)   65 | 8 0 1   74 | 7 0 1   71 | 9 9 4

Divide:

	a	b	c

1) 70 | 8 0 7   81 | 9 2 7   82 | 6 0 7

2) 80 | 9 6 0   81 | 4 0 7   82 | 5 0 7

3) 80 | 8 4 6   81 | 7 4 6   82 | 6 3 6

4) 80 | 5 0 1   81 | 8 0 1   82 | 6 5 7

5) 80 | 9 0 1   81 | 9 0 1   82 | 9 0 2

Divide:

	a	b	c
1)	33 ⟌ 7 9 2 5	32 ⟌ 9 0 7 6	34 ⟌ 9 0 7 0
2)	33 ⟌ 9 1 5 4	35 ⟌ 9 0 7 7	35 ⟌ 9 0 7 1
3)	36 ⟌ 5 4 6 3	36 ⟌ 7 4 6 8	
4)	41 ⟌ 4 0 1 5	42 ⟌ 3 0 1 9	

Divide:

	a	b	c

1) 　　37 | 8 2 7 4　　　　37 | 9 2 7 0　　　　39 | 9 0 9 4

2) 　　39 | 6 1 5 3　　　　39 | 7 0 7 5　　　　39 | 4 0 7 5

3) 　　37 | 5 4 6 4　　　　37 | 5 4 6 3

4) 　　38 | 4 2 1 0　　　　38 | 3 3 1 4

Divide:

	a	b	c
1)	21 ⟌ 8 0 7	22 ⟌ 9 2 7	23 ⟌ 6 0 7
2)	24 ⟌ 9 1 5 0	25 ⟌ 4 0 7 4	26 ⟌ 5 0 7 2
3)	27 ⟌ 8 4 6 1	28 ⟌ 7 4 6 5	
4)	11 ⟌ 4 0 1 2	12 ⟌ 3 0 1 6	

Divide:

	a	b	c

1)

$29 \overline{) 6\ 3\ 6\ 8}$ $14 \overline{) 4\ 0\ 1\ 3}$ $15 \overline{) 3\ 0\ 1\ 7}$

2)

$13 \overline{) 2\ 0\ 4\ 9}$ $31 \overline{) 4\ 0\ 7\ 3}$ $32 \overline{) 5\ 0\ 7\ 8}$

3)

$16 \overline{) 2\ 0\ 5\ 6}$ $31 \overline{) 8\ 4\ 6\ 4}$

4)

$30 \overline{) 4\ 0\ 8\ 4}$ $31 \overline{) 4\ 0\ 1\ 6}$

Divide:

                        a   b   c

1)   30 | 4 0 3 5   14 | 4 0 1 3   15 | 3 0 1 4

2)   66 | 9 0 5 7   71 | 9 0 9 9   75 | 8 2 5 7

3)   72 | 7 9 6 8   73 | 8 4 6 4

4)   75 | 4 0 1 5   71 | 3 0 1 2

Multiplication Worksheets

1 × 0 = 0	2 × 0 = 0	3 × 0 = 0	4 × 0 = 0	5 × 0 = 0
1 × 1 = 1	2 × 1 = 2	3 × 1 = 3	4 × 1 = 4	5 × 1 = 5
1 × 2 = 2	2 × 2 = 4	3 × 2 = 6	4 × 2 = 8	5 × 2 = 10
1 × 3 = 3	2 × 3 = 6	3 × 3 = 9	4 × 3 = 12	5 × 3 = 15
1 × 4 = 4	2 × 4 = 8	3 × 4 = 12	4× 4 = 16	5 × 4 = 20
1 × 5 = 5	2 × 5 = 10	3 × 5 = 15	4 × 5 = 20	5 × 5 = 25
1 × 6 = 6	2 × 6 = 12	3 × 6 = 18	4 × 6 = 24	5 × 6 = 30
1 × 7 = 7	2 × 7 = 14	3 × 7 = 21	4 × 7 = 28	5 × 7 = 35
1 × 8 = 8	2 × 8 = 16	3 × 8 = 24	4 × 8 = 32	5 × 8 = 40
1 × 9 = 9	2 × 9 = 18	3 × 9 = 27	4 × 9 = 36	5 × 9 = 45
1 × 10 =10	2 × 10 =20	3 × 10 = 30	4 × 10 = 40	5 × 10 =50
1 × 11 = 11	2 × 11= 22	3 × 11 = 33	4 × 11= 44	5 × 11 = 55
1 × 12 =12	2 × 12 =24	3 × 12 = 36	4 × 12 = 48	5 × 12= 60

6 × 0 = 0	7 × 0 = 0	8 × 0 = 0	9 × 0 = 0	10 × 0 = 0
6 × 1 = 6	7 × 1 = 7	8 × 1 = 8	9 × 1 = 9	10 × 1 = 10
6 × 2 = 12	7 × 2 = 14	8 × 2 = 16	9 × 2 = 18	10 × 2 = 20
6 × 3 = 18	7 × 3 = 21	8 × 3 = 24	9 × 3 = 27	10 × 3 = 30
6 × 4 = 24	7 × 4 = 28	8 × 4 = 32	9 × 4 = 36	10 × 4 = 40
6 × 5 = 30	7 × 5 = 35	8 × 5 = 40	9 × 5 = 45	10 × 5 = 50
6 × 6 = 36	7 × 6 = 42	8 × 6 = 48	9 × 6 = 54	10 × 6 = 60
6 × 7 = 42	7× 7 = 49	8 × 7 = 56	9 × 7 = 63	10 × 7 = 70
6 × 8 = 48	7 × 8 = 56	8 × 8 = 64	9 × 8 = 72	10 × 8 = 80
6 × 9 = 54	7 × 9= 63	8 × 9 = 72	9 × 9 = 81	10 × 9 = 90
6 × 10 = 60	7 × 10 =70	8 × 10 = 80	9 × 10 =90	10 × 10 =100
6 × 11= 66	7 × 11= 77	8 × 11 = 88	9 × 11 = 99	10× 11 = 110
6 × 12 = 72	7× 12= 84	8 × 12 = 96	9 × 12 =108	10× 12=120

PROGRESS NOTES

www.ingramcontent.com/pod-product-compliance
Lightning Source LLC
Chambersburg PA
CBHW081210180526
45170CB00006B/2294